本书由浦东新区科普项目资金资助

中国珍稀物种科普丛书

娃娃鱼的故事

叶晓青　徐骆羿　著

王　紫　绘

孙博韬　译

上海科学技术出版社

图书在版编目（CIP）数据

娃娃鱼的故事 ：汉英对照 / 叶晓青，徐骆羿著 ；孙博韬译 ；王紫绘. -- 上海 ：上海科学技术出版社，2020.10
（中国珍稀物种科普丛书）
ISBN 978-7-5478-5127-2

Ⅰ. ①娃… Ⅱ. ①叶… ②徐… ③孙… ④王… Ⅲ. ①大鲵－少儿读物－汉、英 Ⅳ. ①Q959.5-49

中国版本图书馆CIP数据核字(2020)第202350号

扫码，观赏“中国珍稀物种”系列纪录片《中国大鲵》

它是世外桃源的隐遁者，它是古老传说中我们祖先的图腾，它拥有一个耳熟能详的名字，可真正能亲眼看到它真面目的人却越来越少。跟随“中国珍稀物种”系列纪录片《中国大鲵》走进三亿年前的远古精灵。影片荣获国家科技进步二等奖、上海市科技进步一等奖、星光奖等奖项。

中国珍稀物种科普丛书
娃娃鱼的故事

叶晓青　徐骆羿　著
王　紫　绘
孙博韬　译

上海世纪出版（集团）有限公司
上 海 科 学 技 术 出 版 社　出版、发行
（上海钦州南路 71 号　邮政编码 200235　www.sstp.cn）
浙江新华印刷技术有限公司印刷
开本 889×1194　1/16　印张 4
字数：70 千字
2020 年 10 月第 1 版　2020 年 10 月第 1 次印刷
ISBN 978-7-5478-5127-2/N·212
定价：50.00 元

导读

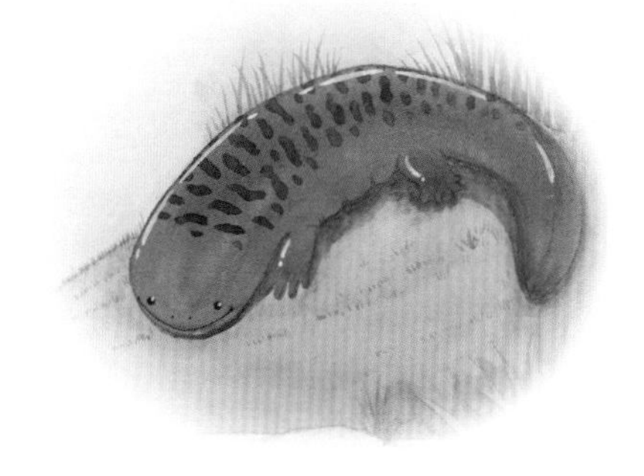

娃娃鱼保护中心新出生了很多小娃娃鱼，阿宝是其中最强壮的一条。它在保护中心长大，直到有一天，它被放归到野外。它第一次感受到太阳的灼热，第一次遭遇到风餐露宿的窘境，第一次躲避向它发起攻击的同类……阿宝觉得很迷茫，大自然到底好在哪里？

本书分为上下两个部分。第一部分采用儿童喜闻乐见的绘本故事形式，在尊重科学事实的基础上，将充满趣味的故事与精美的绘画相结合，提升整体艺术表现力，给读者文字以外的另一个想象空间。第二部分采用问答的形式，增进公众对该珍稀物种的科学认识，通俗易懂的语言，配上精美的照片，有利于儿童的阅读和理解。是一本兼具科学意趣和艺术质感的少儿科普读物。

目录

我不喜欢大自然

我叫阿宝，我从出生就住在这个“小房子”里。我的房子很特别，墙壁好像是一层软绵绵的磨砂玻璃，很有弹性，远远看上去像个不规则的水晶球。这栋房子好像有魔法，给我源源不断的能量。

My name is Po. I was born in this little house, which looked like an irregular crystal ball. The stretchy wall seemed to be a layer of soft ground glass. The house kept providing me energy magically.

这是哪里？我的家吗？

Where is this place? My home?

我对一切都充满好奇，没事的时候喜欢打量四周，放眼望去都是我的兄弟姐妹，我们的房子像珍珠项链一样串成一串。有的房子是独成一室的，有的是两居室，甚至三居室，特别热闹。我们现在只能隔着墙打招呼，不知道什么时候能出去一起玩呢？

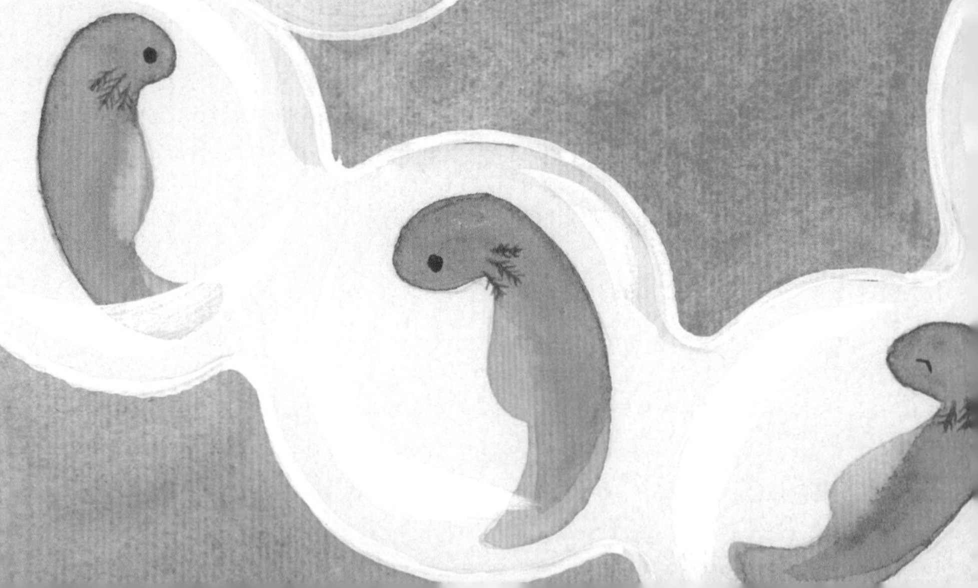

I was curious about everything. I liked to look around and see my brothers and sisters everywhere. Our houses were strung together like pearl necklace. Some houses had one room, some had two or three. We could only greet to each other across the wall. When could we play together?

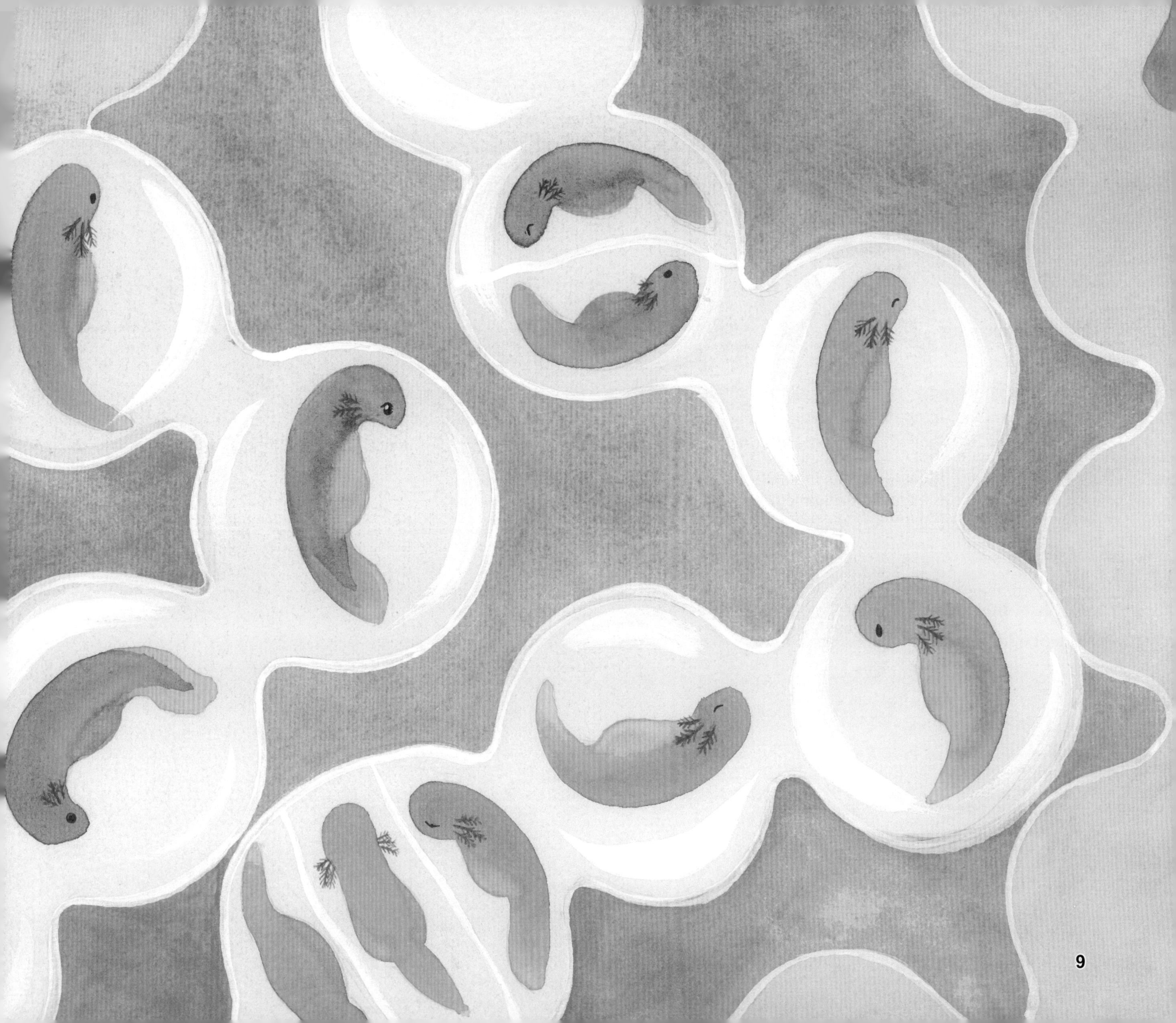

我们从小就没见过妈妈，平时都是爸爸照顾我们，他将长长的身体盘成半圆形，我们就像被他抱在怀里。我们在一天天长大，每天都有穿白大褂的人来看望我们，他们还会拿着一张纸写各种数字。

有一天，爸爸对我们说："孩子们，已经快40天啦，你们马上就要出生了。"

We haven't seen mom since we were babies. Dad usually took care of us all the time. He twisted his long body into a semicircle to protect us. People in white coats came to see us every day. They recorded the numbers on paper.
One day, Dad told us: "My children, it has been nearly forty days. Soon you'll be born."

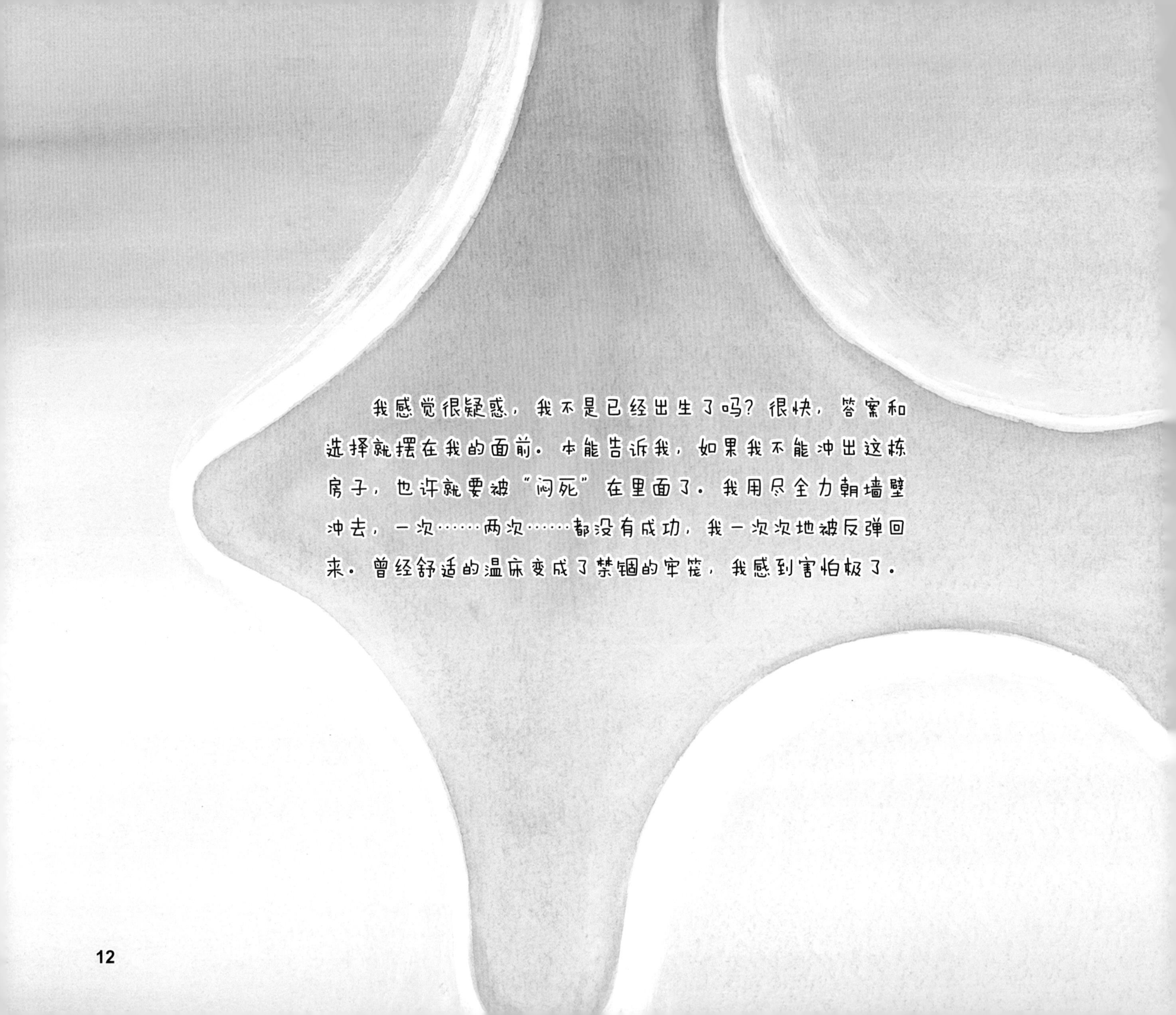

我感觉很疑惑，我不是已经出生了吗？很快，答案和选择就摆在我的面前。本能告诉我，如果我不能冲出这栋房子，也许就要被“闷死”在里面了。我用尽全力朝墙壁冲去，一次……两次……都没有成功，我一次次地被反弹回来。曾经舒适的温床变成了禁锢的牢笼，我感到害怕极了。

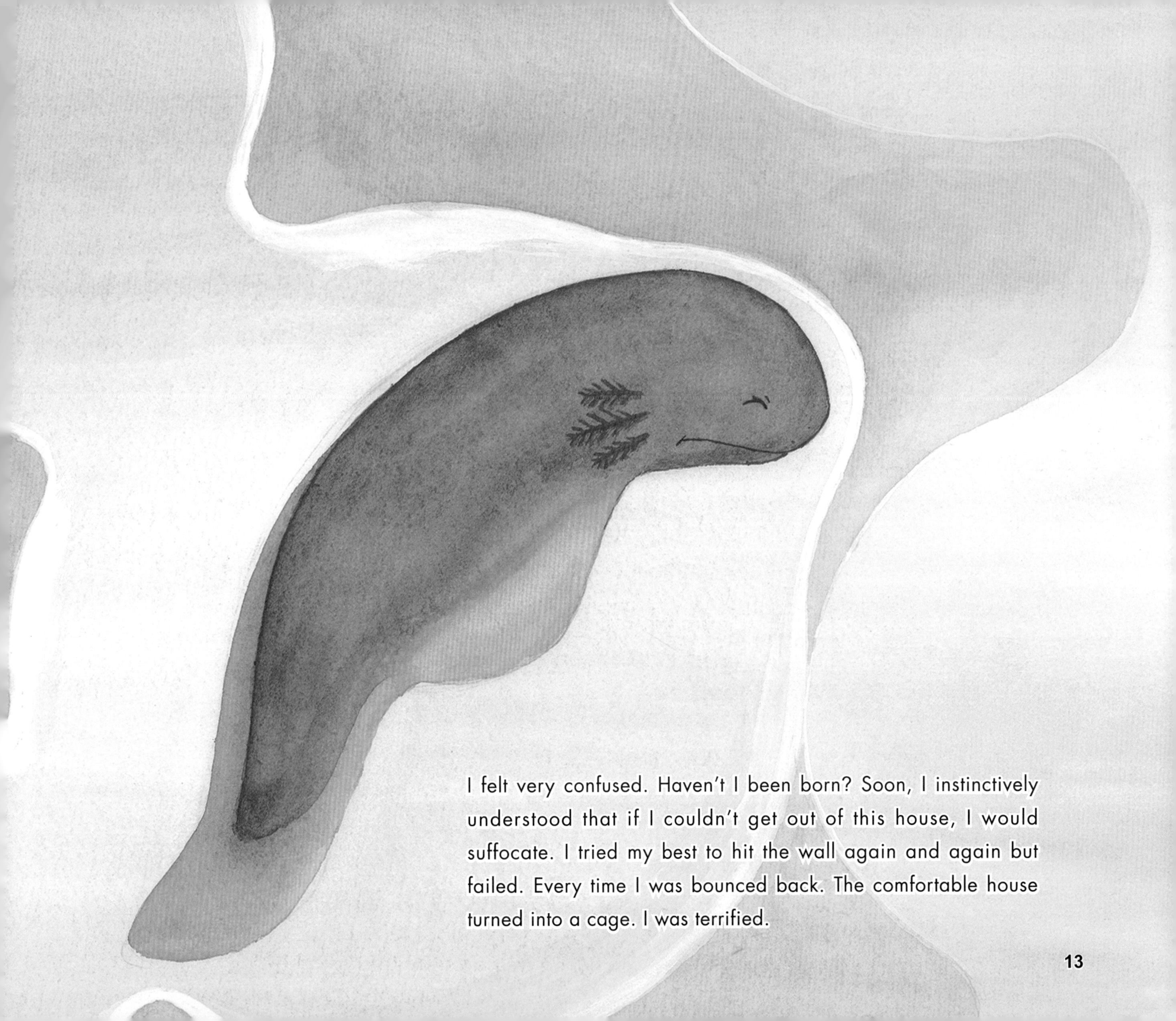

I felt very confused. Haven't I been born? Soon, I instinctively understood that if I couldn't get out of this house, I would suffocate. I tried my best to hit the wall again and again but failed. Every time I was bounced back. The comfortable house turned into a cage. I was terrified.

我能感受到爸爸的焦急，听到他不停地为我们打气。我有些脱力，但不能放弃。我用力晃动着脑袋，摆动着细小的尾巴。我感觉到一直撞击的某一点变脆弱了。最后一击，我从房子里一跃而出！成功了！我开心地甩起小尾巴，赶紧向爸爸身边扑腾过去。我边游边想，一定要记住今天的日子，今天应该算是我真正的生日吧？

I could feel dad's anxiety and hear him keep cheering us up. I was very tired, but I couldn't give up. I shook my head and wiggled my little tail. I felt that a point on the wall had been weaker. With my last try, I successfully jumped out of the house! I swam to dad happily. Today must be my real birthday, I thought.

别放弃，加油啊，孩子！
Don't give up! Go for it!
爸爸，爸爸！我成功啦！
Dad! I did it!

“爸爸，这是哪里？这是我们的家吗？”

“这是人类为我们建的家园。”

我圆溜溜的小眼睛向四周不断地张望，这里比我原来的小房子大太多了，感觉都游不到尽头。

人类为我们建的家园？我有些似懂非懂。爸爸说，这里是娃娃鱼保护中心，他和妈妈是从野外被救助到这里的，而我们是在这里出生的第三批孩子。原来那些穿着白大褂的人是这里的科研人员，自我们出生起他们每天都来记录我们的生长数据。

“Dad, where is this place? Our home?”

“This is our home. Humans built it for us.”

I looked around. This place was much larger than my previous house. I couldn’t swim to the end.

A home built by humans? I didn’t understand. Father told me that this was Giant Salamander Protection Center. He and my mom were rescued from the wild. My brothers and sisters and I were the third generation of giant salamander born here. People in white coats were scientific researchers. They recorded our growth data every day since we were born.

娃娃鱼保护中心

我一直有个大大的疑问，为什么我和爸爸长得完全不一样呢？他有一张大大的嘴，一个宽宽扁扁的大脑袋，相比之下小眼睛显得有些滑稽，不过小小的眼睛里透出睿智的光芒。爸爸的身体上还有像花边一样的褶皱，游动的时候能划出一道道美丽的水波。短短的四肢像是4个小胳膊，尾巴大概有身体的三分之一长。再看看我，连四肢都没有，头上两边还各长着3根外鳃，和小鱼差不多。

I always had a big question, why did I look different from dad? He had a big mouth. Comparing to his broad and flat head, his small eyes were very comical. However, there was a light of wisdom in his little eyes. There were lace-like wrinkles on his body, which could draw beautiful waves when he swam. His limbs were short, and the tail was about one-third length of the body. Look at me, I didn't have limbs. I only had three external gills on both side of my head. I looked like a fish.

爸爸，为什么我长得不像你啊？
Dad, why don't I look like you?
再等一段时间，你就知道啦！
Wait for some days and you'll know.

“我的手呢？会慢慢长出来吗？”

我晃晃小脑袋，还是有些不明白。

我之所以能在水里呼吸，就是依靠脑袋上像树枝一样的外鳃，它们看上去还毛茸茸的。

“我的脚呢？会慢慢长出来吗？我的皮肤会变成爸爸那种颜色吗？爸爸头上为什么没有外鳃呢？”我看着自己光溜溜的身体，感到很疑惑。

爸爸偶尔会离开水面，而我现在连游泳都很困难，只能侧着身体贴在水底。“水面外的世界是什么样子的呢？”

"Where is my hand? Will they grow in the future?"
I swung my head, still didn't understand.
The reason why I could breathe in the water was that I had the external gills, which looked like branches on my head. They looked fluffy.
"Where is my feet? Will they grow? Will my skin turn to color of dad? Why dad has no external gills?" I felt confused about my smooth body.
Father occasionally left water. I even couldn't swim well. Thus, I could only lie on the bottom of water. "How did the world outside water look like?"

突然有一天，我惊奇地感觉到身体的异样，两条前腿居然慢慢长出来了。又过了一段时间，后腿居然也长出来了一点点。虽然还都是肢芽的样子，但我已经迫不及待了。

One day, I was surprised to find that my two fore limbs grew out.
Two hind limbs then appeared. I couldn't wait for them.

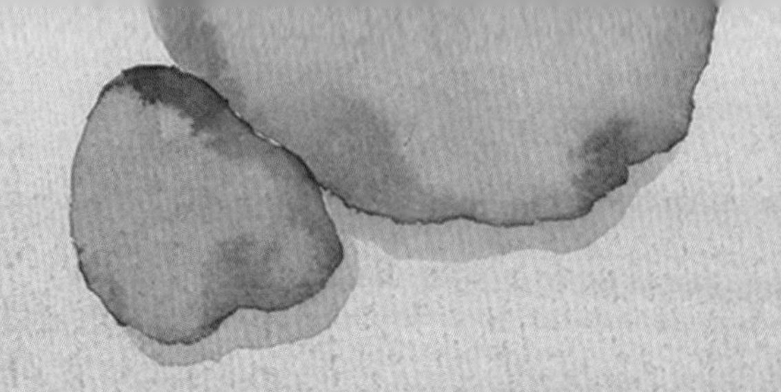

我的四肢越来越有力，好几次居然很顺利地就能把身体撑起来在水中滑行一段距离了呢。

“咦?！我有脚趾头了！”好像发现了新大陆一样，我低头看到我的四肢分叉成了好几个脚趾头。

“一、二、三、四，一、二、三、四、五，我脚趾的数量不一样啊！”本来还很开心地数脚趾头的我，被吓了一大跳，哇哇大哭起来。

爸爸很淡定，“阿宝，别担心，我们前后脚趾的数量本来就不一样哦。你看，我们的前肢是4个趾头，后肢是5个趾头。我们这几个趾头和人类刚刚出生的婴儿的小手很像，这也是我们被称为‘娃娃鱼’的原因。”

My limbs were getting stronger. Sometimes I could support my body to glide in water.
I have toes! Toes appeared at the end of my limbs.
“One, two, three, four. One, two, three, four, five. The number of my fore toe and hind toe is different!”
I was taken aback and cried.
Father was calm, “Po, don’t worry. The number of our fore toe and hind toe is indeed different. Look, we have four fore toes and five hind toes. Our toes are similar to human baby’s hand. That’s why human call us ‘baby fish’.”

哎呀！爸爸！出大事了！我前后脚趾头的数量怎么不一样啊？怎么办？我是不是生病了？

Ah! Father! Why the numbers of my fore toe and hind toe are different? Am I sick?

脑袋上的外鳃开始萎缩，甚至脱落了，我有点慌张，如果没有外鳃，我怎么呼吸啊？

今天，是我出生后的第9个月，爸爸说要带我去一个地方。他越游越远，居然爬到了岸上，"阿宝，到爸爸这里来。"

我使劲摇头："不行的，我只能在水里呼吸，离开水，我会死的。"

爸爸看着我，温柔而坚定地说："阿宝别怕，你长大了，这是必经的过程。"

我的身体僵在水里，这一步虽然很小，但对我来说却是最艰难的一步。我想起小时候看着爸爸上岸的背影，心底有一个声音让我勇敢一点。

我小心翼翼地从水里探出脑袋，小得像针眼一样的鼻孔暴露在空气中，一动不敢动。直到憋得胸口发疼，才迫不得已试着慢慢用鼻子吸了一口空气。这感觉就像是开启了自动马达，身体自然而然就能呼吸了。我才意识到这几个月变化的不仅仅是我的脚，还有我身体里的某个器官。

My external gills atrophied, then fell off. How could I breath without the external gills? I was a little flustered.

It had been nine months since I was born. Father said he was taking me somewhere. He swam farther and farther until he was on the bank. "Come to me, Po."

"No. I can only breath in water. I'll die without water!" I shook my head.

"Don't be afraid, Po. It's part of growing up." Father said tenderly and firmly.

I was numb in water. This small step was too difficult for me. I remembered when I was a child watching my father on the bank. A voice in my heart told me to be brave.

I carefully poked my head out of the water. My little nares were exposed to the air. I dared not move. Then I tried to breathe in slowly through my nose. Finally, I learnt to breath naturally. I realized that not only my limbs and toes, but also an organ had changed.

“那是你的肺在呼吸。”爸爸一眼就看穿我的疑惑，“小时候你在水里只能用外鳃，而现在你的肺已经成熟了，以后就主要靠肺来维持呼吸了。这个过程对我们两栖类很重要。”

陆地就在我的眼前，我用小小的脚试探地去碰了碰岸上的土地，这是一个全新的体验。

“That’s your lungs breathing.” Father knew my confusion, “When you were a hatchling, you could only breath by external gills. Now your lungs are mature. You can breathe by them. This is important for amphibians.”
Land was in front of me. I touched the land tentatively. It was a new experience.

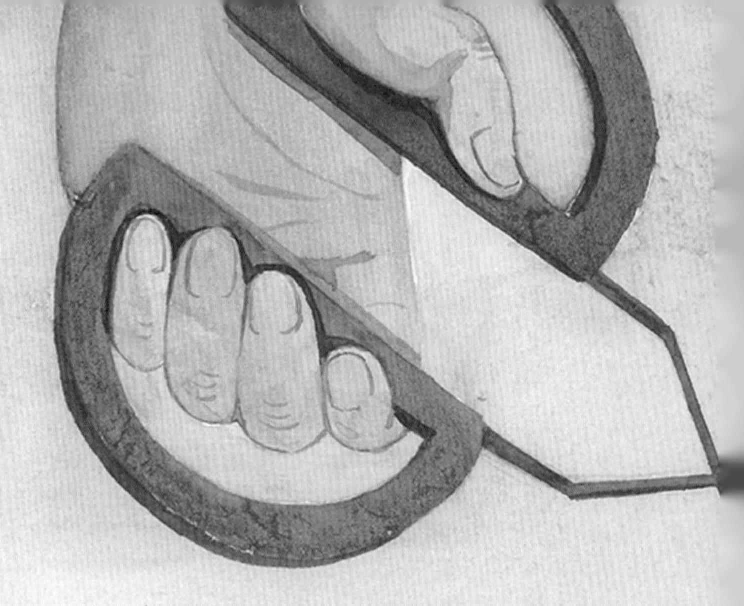

又过了3个月，除了个头，我长得和爸爸几乎一模一样了。我已经能比较熟练地往来于岸上和水下。冬天到了，虽然室内还比较温暖，但爸爸带着我们一起进入了冬眠。

时间过得很快，分离总是来得猝不及防。渐渐长大的我被带离了爸爸的身边，转移到了旁边的池子里。

我吃的东西也开始发生变化，从以前的小虾、小昆虫，到现在的小鱼，吃的东西越来越多。第一次吃鱼的时候，我根本就吃不到，小鱼游来游去，我还以为是放在池子里陪我玩的。直到科研人员重新把鱼捞起来，再用长钳子夹住放在我的嘴边，我才明白了他们的意思，张口把鱼咬住吞了下去。

Three months later, expect for my size, I looked almost exactly like my father. I was able to crawl between bank and water deftly. Winter had come. Although it was relatively warm indoor, father guided us into hibernation.

Time flies. Unexpected separation had come. As I grew up, I was taken away from my father and moved to the next pool.

My food changed as well. I ate shrimps and insects before. Now fish had been added. I couldn't catch a fish when I saw them for the first time. I thought they were here to play with me. Until a researcher clamped a fish using a tweezers and put it near my mouth. Then I understood that fish was my food and ate one.

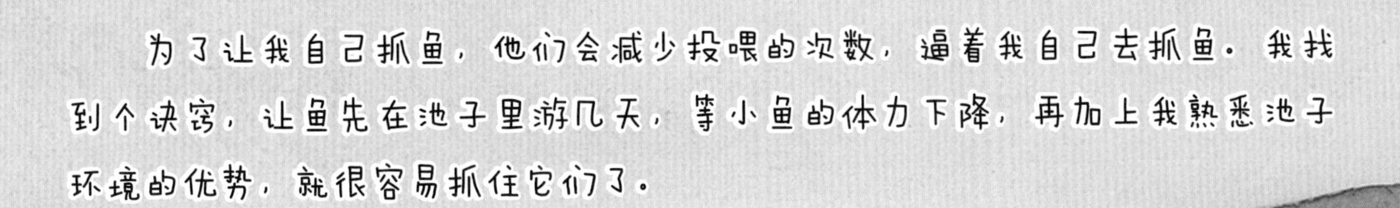

为了让我自己抓鱼，他们会减少投喂的次数，逼着我自己去抓鱼。我找到个诀窍，让鱼先在池子里游几天，等小鱼的体力下降，再加上我熟悉池子环境的优势，就很容易抓住它们了。

They reduced the feeding times to force me to catch fish myself. I had the advantage of being familiar with the pool environment. Let the fish swim in the pool for a few days. When the fish's strength failed, it was easy to catch them.

科研人员还是每天来检查和记录我的身体情况，他们说我是这批新出生的娃娃鱼里最强壮的一条。听说他们有个计划，送我回到我该去的地方。

那个地方叫做大自然。

我看科研人员忙忙碌碌的，还听到他们说已经考察了很多地方，最后选定了一个最为合适的。

大自然对我来说很陌生，只在小时候爸爸对我说过。我很茫然，为什么我要去那里，保护中心不是很好吗?

Researchers still came to inspect and record my physical condition every day. They said I was the strongest one among all hatchlings of that year. I heard they have a plan to get me back to where I belong.
That place is called Nature.
Researchers were busy. I heard that they had investigated many places and chosen a most appropriate one.
I was not familiar to nature, though father had talked it to me in my childhood. I was confused. Why should I go to nature? Wasn't Protection Center good?

00.0 kg

临出发前，爸爸说他为我能有机会回到大自然感到特别开心，那里才是我们真正的家。

经过很长的车程颠簸和徒步行走，我被小心翼翼地放在深山中的一条小溪里，冰凉的溪水让我打了个哆嗦。这是我第一次见到外面的世界，一切都很新鲜。

当我还在睡觉的时候，头顶的光线越来越强，刺得我眼睛发疼。我的脑袋一露出水面，就被炙热的阳光照得皮肤有种被灼伤的感觉，吓得我连忙躲回水中，摸索着朝岸边躲去。保护中心的光线一直都是比较暗的，那是为了适应我们夜行的习惯。可是现在，我只能缩在石头缝里一动不敢动，焦急地等待太阳下山。

Before leaving, father said he was very happy for me to get back to nature. That was our real home.

I was carefully placed in a stream deep in the mountains after a long journey. Icy stream made me thrilled. This was the first time to touch the world outside the Protection Center. Everything was new for me.

The sunlight became blazing when I was asleep. I felt burned when my head was above the water. I was frightened that I quickly hid back into water. I groped my way to the shore. In Protection Center, it was always dusky to adapt our nocturnal habit. Now I could only hide between the stones, waiting for night.

走开，小个子，这个地方是我的。
Get out of here, you little one. This is my place.
可是……我走了很长的时间了……我想休息一下……
But... I crawled too long... I only wish to have a rest...

直到月亮出来了，我才慢慢爬出来，舒展一下蜷缩了一天的身体。我必须要先找到一个栖身之地，我可不想天天住在石头缝里。

趁着月光，我顺着小溪一路游走，一路找，就在我快筋疲力尽的时候居然很幸运地发现了一个洞穴。

突然冲出来一个大黑影，朝我顶了过来，把我给掀翻了。我稳住身体，定睛一看，是一条又壮又凶悍的娃娃鱼，他朝我呲着牙，不准我进入他的地盘，粗壮的尾巴甩在我身上真疼。

眼看天又要亮了，我只能再次找一个石头缝躲起来。

I slowly crawled out and stretched out my body after a long day until the moon came out. I had to find somewhere to live. I don't want to live between stones every day.

I swam down the stream, seeking for a shelter. Fortunately, I found a cave before getting exhausted.

A big shadow suddenly rushed out and overturned me. It was a strong and fierce giant salamander. He prevented me from entering his territory. His strong tail swung against me. That hurts!

Dawn was to come. I had to hide between stones again.

“啊！有怪物！”

我洞穴没有想象中容易，一连找了好几天，都没有合适的。前天下午我在睡觉，有一根树枝伸到了石头缝里，眼看就要戳到我了，我“嗖”地一下从石头缝里蹿了出来，就听到一声尖叫。

我抬头看到岸上有一个五六岁的小孩，用树枝指着我的方向。

“怪物？我是娃娃鱼！”我有点生气，朝着小孩张开嘴。

小孩以为我要咬他，居然捡起石头朝我扔了过来。

“咚！”石头落在我身边，在水中激起了一个大水花。吓得我赶紧往对岸游，继续找个石头缝躲起来。

“A monster!”

It was not easy to find a cave. I had been looking for it for days, but nothing worked. One day, a branch stuck into the stone and nearly stabbed me. I rushed out and heard a scream.

I looked up and saw a child five or six years old on the bank, pointing my direction with a branch.

“A monster? I’m a giant salamander!” I was somewhat angry. I opened my mouth towards him.

That human child thought I was trying to bite him. He picked up a stone and threw it at me.

The stone fell beside me, causing a big splash. I hurried to swim to the other side of the stream and hid in another cleft.

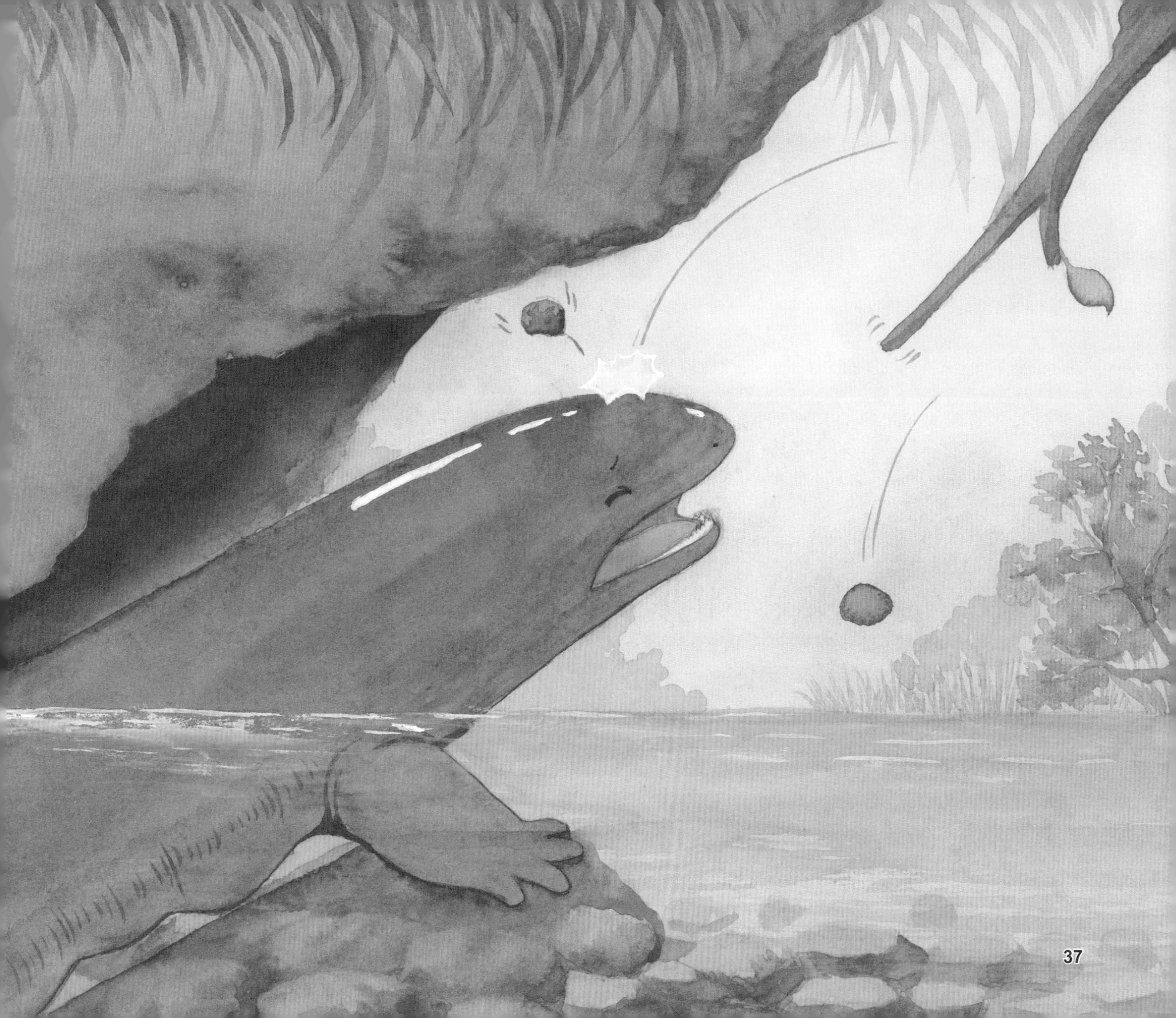

经过几天的折腾，我又累又饿。晚上忙着赶路，根本就没时间吃饭。小溪里不乏小鱼、小虾，可我的视力不好，只能靠水的波动来判断食物在哪里。

左边有一条小鱼，我朝前一扑，没扑到。

右边有一群小虾，我朝前一咬，没咬到。

试了几十次，都是竹篮打水一场空，我的肚子反而更饿了。

这溪水里的小鱼、小虾也太灵活了，一转眼就找不到了。大自然一点都没意思！真想念保护中心，有住的地方，有好吃的东西，有合适的光线，还有人来检查身体。为什么要把我放到大自然来？我越想越委屈，哇哇大哭起来。

I was tired and hungry after several days. I didn't have time for hunting as I need to travel at night. There were no lack of small shrimps and fish in the stream. However, my eyesight was poor, I could only locate food according to water wave.
Lots of shrimps and fish were around me. I bit but missed.
I tried dozens of times and got nothing. I felt even more hungry.
Fish and shrimps here were too agile to catch. Nature was so boring! I missed the Protection Center so much. I missed the comfortable place, the delicious food, the proper light, and researchers' examination. Why did they send me to nature? I felt wronged and started crying.

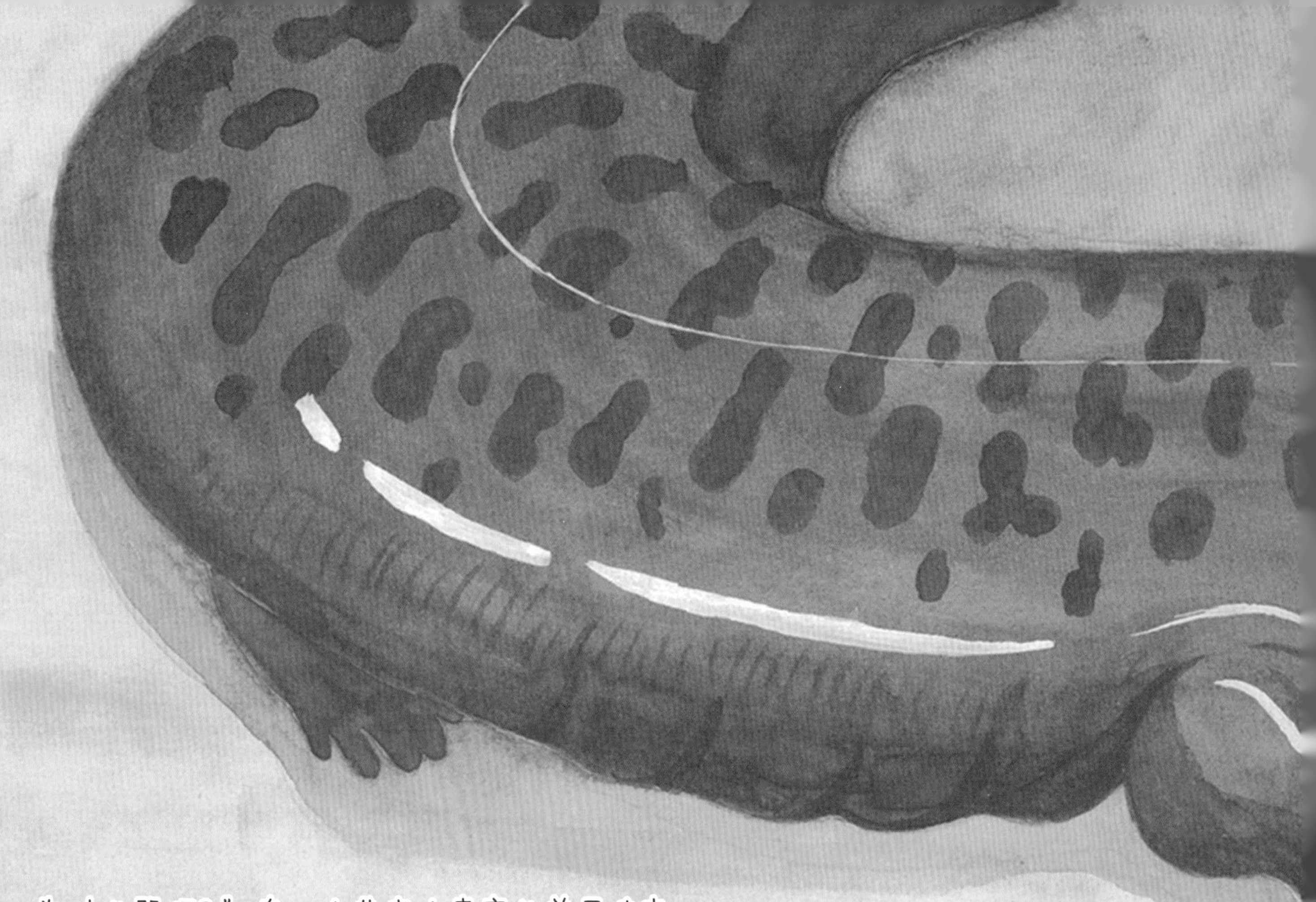

“小家伙，为什么哭啊？”有一个苍老的声音从前面传来。

天呐？这是？抬头一看，我一下子就愣住了，惊讶地停止了抽泣。

这是一条比我大4倍的娃娃鱼，他的身体大概有2米了吧？这简直就是童话世界里的巨人。

我战战兢兢地老实回答：“我没有住的地方，也饿了好久，我想回家。”

我的注意力又被鲵爷爷有些与众不同的肤色吸引，“您身体的颜色和我不太一样呢。”

“我身上是不是有点泛红？小的时候颜色还要漂亮，我们娃娃鱼颜色可是多种多样的哦。”鲵爷爷笑呵呵地说道：“小家伙，别怕，你先和我回家住一段时间吧。”

"Little one, why are you crying?" I heard an old voice.
Wow, is this...? I froze and stopped sobbing in surprise.
That was a giant salamander four times bigger than me. He was about two meters long. A real giant.
I answered tremblingly, "I have no place to live and have been hungry for a long time. I want to go home."

I was attracted by the unique skin color of him, "The color of your body is not the same as mine."
"Yes, I'm red. I was more beautiful when I was young. Skin color of us giant salamanders varies. Don't be scared, little one. Go home with me for a good rest."

我觉得这个世界太神奇了，一路上像只小麻雀似地叽叽喳喳问个不停。

“爷爷，我叫阿宝，你一直都生活在这里吗？”

“爷爷，你怎么这么大啊？”

“爷爷，这里的鱼、虾怎么这么难抓啊？”

鲵爷爷笑呵呵地听着我的各种各样的问题，开口道：“我已经150岁喽，所以身体那么大呀。”

“150岁？”我大吃一惊。

“是啊，我从小就生活在这个流域，以前同伴还很常见，慢慢地越来越少了。后来我搬到了这条小溪里，这里人少，环境也好，不受打扰。我们娃娃鱼的自然寿命长着呢。”鲵爷爷很自豪地说。

The world was amazing. I kept talking along the way.

“Grandpa, my name is Po. Have you always lived here?”

“Grandpa, why are you so big?”

“Grandpa, it's too difficult to catch fish and shrimps here.”

Grandpa Salamander laughed and answered, “I'm one hundred and fifty years old. Thus, I'm so big.”

“One hundred and fifty years old?” I was surprised.

“Yes. I have lived in this basin since I was a hatchling. My companions, which used to be common, have gradually become less and less. Then I moved to this stream. The environment here is good and quiet. Nature lifespan of us giant salamander is very long.” Grandpa Salamander said proudly.

夜幕降临，热闹了一天的山林渐渐归于沉寂，习习夜风把白天的热气轻轻吹散，皎洁的月亮静静挂在天上，为黑夜点起了一盏灯。

鲵爷爷把我带到洞口水边的乱石丛里，用身上与石块相近的保护色做掩护。“抓鱼的关键就是耐心、迅速。追着鱼走，那成功率是很低的哟。”

等到水面平静下来，小鱼慢慢靠近。说时迟那时快，鲵爷爷突然发动袭击，用尖利的牙齿死死咬住猎物，然后囫囵吞了下去。

我有模有样地学着鲵爷爷的样子开始伏击，一开始都以失败告终，后来技巧越来越娴熟，一个晚上下来，居然抓住了好几条鱼。

Tranquility returned to nature when the night fell. Cool breeze blew away the heat of the day. The bright moon hung quietly in the sky, lighting a lamp for the night.

Grandpa Salamander brought me to the entrance of the cave, screening themselves with stones that had similar color to them, “The key is patience and swiftness. Chasing a fish is not a good idea.”

Fish got close to us as water became calm once again. Grandpa Salamander ambushed suddenly, sinking his teeth into a fish. Then he swallowed it.

I learnt to hunt. It was difficult at the beginning. Then I mastered the skill. I caught several fish at that night.

又是一个晚上，漫天的繁星点缀在黑丝绒般的天空中。

"都说大自然是我应该归属的地方，可我一直觉得保护中心更好，有吃的，还很安全。"我很怀念保护中心。

鼩爷爷没有说话，半晌才悠悠地说道："阿宝，你仔细看过这个世界吗？等你静静地去感受一下风，感受一下水，感受一下大自然，你就会知道，为什么我们属于这里。"

我仰起头，徐徐清风从身上拂过，是冷冽而清新的气息；叮咚的山泉在脚边潺潺流淌，是甘甜而纯净的味道；摇曳的花儿漫山遍野肆意绽放，是明媚而娇艳的模样；林间鸟儿和松鼠嬉笑打闹，充盈着安逸而美好的感动。

Another night fell. The black velvety sky was full of stars.

"It is said that I belong to nature, but I always prefer the Protection Center. There was enough food, and it was safe." I missed the Protection Center so much.

Grandpa Salamander answered after a while, "Po, have you carefully observed this world? Feel the wind, water, and nature. Then you'll know why we belong to here."

I looked up,

Cool breeze caressed me, bringing freshness.

Mountain springs flew gently, tasting sweet and pure.

Flowers came into bloom all over the hill, looking so bright and charming.

Birds and squirrels were playing on the trees, bringing peaceful and happiness.

我突然有些明白大自然的意义，世间万物，共生共荣，云卷云舒，花开花落，也许这就是大自然给我们的馈赠。

I suddenly understood the meaning of nature. All creatures in the world live in symbiosis, harmony and prosperity. Clouds gather and lift, flowers bloom and fall, maybe this is nature's gift to us.

娃娃鱼的秘密知多少

我的名片

中国大鲵（学名：*Andrias davidianus*）隶属于两栖纲有尾目隐鳃鲵科，国家二级保护动物，列入《华盛顿公约》CITES 附录 I 级保护动物、《世界自然保护联盟濒危物种红色名录》（IUCN）“极危”（CR）物种。世界上现存的隐鳃鲵科包括亚洲东部的中国大鲵、日本大鲵和美国东部的北美隐鳃鲵，仅产于中国的中国大鲵是其中体型最大的一种。

为什么叫它“娃娃鱼”

大鲵俗称“娃娃鱼”，这个名字的渊源可以追溯到几千年前。战国时期的《山海经》是最早记录大鲵的典籍之一，里面有这样一段描述：“龙侯之山，泱泱之水出焉，东流注于河，其状如人鱼，四足，其声如婴儿。”这部中国最早的地理古籍形象生动地勾勒出了大鲵的模样：它多数时间像鱼一样生活在水中，却比鱼多了四只脚，就好像婴儿的四肢，发出的叫声像极了婴儿的啼哭。所以，在民间“娃娃鱼”的称呼就流传了下来。

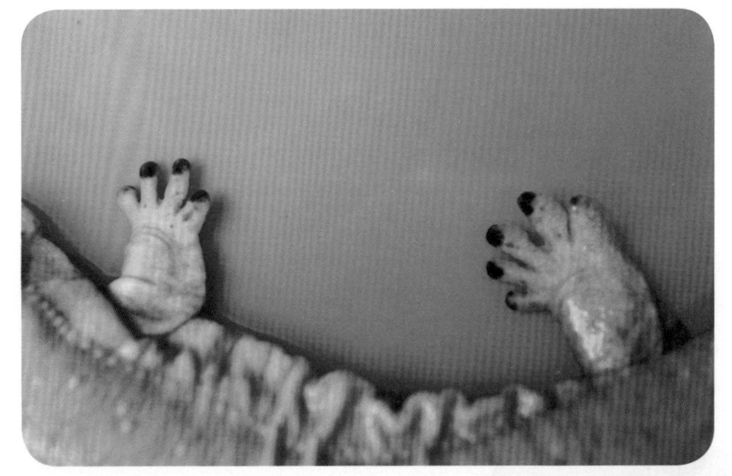

娃娃鱼会叫吗

较之于“四肢外形与人类婴儿相似”这个原因，大多数人更偏向于因为它的叫声像婴儿啼哭所以称它为“娃娃鱼”。在我国《史记》《水经注》等五十多部典籍中都提到了“鲵鱼声如小儿啼”这一特征。然而现代科研人员在对大鲵进行形态学研究时，却惊讶地发现大鲵并没有声带，这也就意味着大鲵即使发出声音，也只可能是危急情况下高速气流通过喉咙发出的声响。我们在科普纪录片《中国大鲵》的拍摄过程中，向有经验的养殖户和周边的居民求证，他们都表示没有听过大鲵发出“如小儿啼”的声音。在影片中，科学家将偶然采集到的大鲵叫声和婴儿哭声进行了声谱比对，发现差异性很大。关于大鲵“发出婴儿啼哭般的叫声”的这一特征仍然是一个谜，所以民间把大鲵唤作“娃娃鱼”的主要原因可能还是在于它的四肢酷似婴儿吧！

为什么娃娃鱼被称为“活化石”

“活化石”指的是一种古老动物或者植物的后代，这种生物千百万年来基本上没有什么显著的进化，就好像它是活着的化石。

两栖类是脊椎动物从水生到陆生进化的一个过渡类群，在地球生命进化研究中占有重要地位。在3个现生两栖动物类群里，青蛙等无尾类的尾巴已经退化了；最低等的蚓螈类已经发生特化；只有大鲵所属的蝾螈类最接近祖先的形态，既有尾巴，也有四肢。大鲵的祖先是比我们人类古老得多的地球“居民”，早在联合体大陆尚未形成漂流大板块之前的泥盆纪，它们就已经在地球上繁衍生息了。大鲵对研究两栖动物的起源和进化有着重要意义，因此被称为生物进化的“活化石”。

娃娃鱼能够长多大

虽然名字里有“娃娃”两个字，但是娃娃鱼可是实打实的大个子。我们人类生长发育完成后，个子就不长了。娃娃鱼可不一样，别看它出生时体长只有28～31.5毫米，随着年龄的增长，它的体长和体重也会直线上升。成年的大鲵可以长到1～1.5米，在栖息环境较好的地方，体长可以长到2米左右。曾经在湖南省桑植县捕到一条长达3米多、重达73.5千克的巨型大鲵。中国大鲵是现生两栖类中体型最大的，被称为“两栖之王”。

娃娃鱼能活到多少岁

自然界中许多动物的寿命都远远超出我们的想象。比如鲤鱼可以活到50岁，鲇鱼可以活到60岁。不少鸟类都能活到40岁以上，其中有些鹦鹉甚至能活到70岁。要说位居前列的长寿动物，爬行动物的寿命可算是令人瞠目结舌了，鳄鱼的寿命可以达到150岁，而龟类更是长寿的代表，甚至有报道称海龟可以活到400～700岁。

生物的寿命长短是有一定规律的。一般而言，体型越大，寿命越长；生长越慢，行动越慢，寿命也越长。在两栖类中，蛙类的寿命就和它的体型成正比，普通的青蛙能活5年，体型巨大的牛蛙就能活16年左右。而娃娃鱼的寿命可以高达130年，甚至150年。因为它是两栖类中体型最大的，行动也很缓慢，同时具备了这两个条件，让娃娃鱼成了两栖类中的长寿明星。

娃娃鱼都长得一样吗

常见的娃娃鱼主要以灰褐色为主，虽然它不会像变色龙那样根据环境来改变颜色，但身体的颜色也会各有不同。我国的很多省市曾经发现过各种颜色的娃娃鱼：湖南省曾经发现过身体通红的娃娃鱼，也出现过全身金黄色的娃娃鱼；四川省发现过白色的娃娃鱼；广东省发现过带黑点的红色娃娃鱼；重庆市发现过棕红色的娃娃鱼。科学家认为，娃娃鱼在身体颜色上的种群变异和它生活的环境有一定的关系。生活在黄河流域，如甘肃省、河南省的娃娃鱼身体颜色偏黄色，斑纹比较少；生活在长江流域，如湖南省、湖北省的娃娃鱼身体颜色偏黑色，斑纹比较多；生活在珠江流域，如广西壮族自治区的娃娃鱼身体颜色为红棕色的比较多。而金黄色、白色的娃娃鱼属于特殊变异，就非常稀有了。

娃娃鱼是一家人一起生活吗

娃娃鱼平时基本上都是独居的，它喜欢独自生活在一个独立的洞穴里面，而且绝不会允许其他同类入侵自己的家。但是到了繁殖期，雄性娃娃鱼会游到雌性娃娃鱼的栖息地，一起寻找一个新的洞穴。这个洞穴就是给小娃娃鱼们生长的洞穴，等到小娃娃鱼们孵化并能独立生活以后，它们会离开洞穴，自己去寻找合适的栖息地。一般来说，娃娃鱼一生中有80%以上的时间都是单独生活的。

娃娃鱼是怎么出生的

娃娃鱼是卵生动物，虽然它们的寿命很长，但繁殖期却很短，只有10年左右。在自然界中，娃娃鱼要长到5～6岁，才具备寻找配偶、繁殖的能力；而到16岁左右，它们便逐渐丧失生育功能，以后便不再进行繁殖了。

娃娃鱼一年繁殖一次。娃娃鱼妈妈一次可以排出400～1 500枚卵。刚生出的卵球看上去接近透明的乳白色，卵径为5～8毫米，中间是乳黄色的卵黄。从内到外有三层膜，分别为

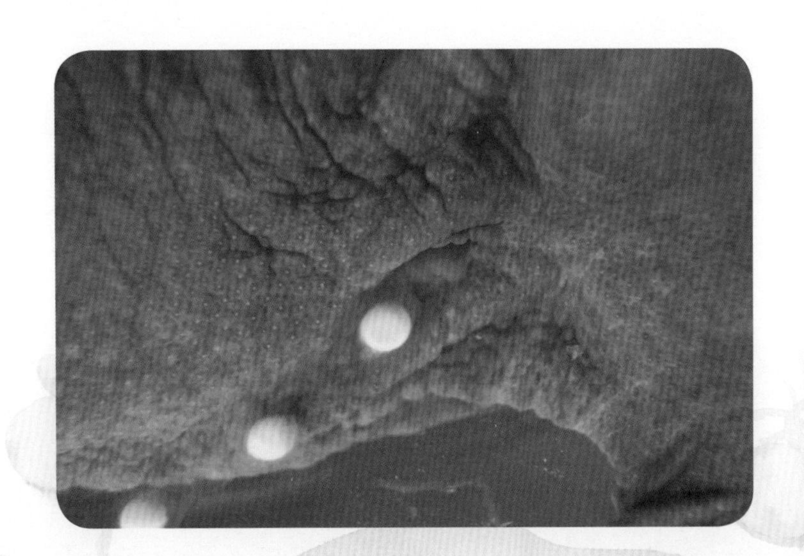

卵黄膜、内胶膜和外胶膜。卵球与卵球之间连在一起，差不多有1～3厘米的间距，这么多的卵球相连就形成了长达数米至数十米的念珠状卵带。卵带飘浮在水中，有时也会粘在石壁或者挂在水草上，在阳光的照耀下就像一条晶莹剔透的珍珠项链。

娃娃鱼是妈妈照顾长大的吗

在自然界中，大多数的哺乳动物在出生后，都是由妈妈照顾长大；90%以上的鸟类的育雏工作是由父母共同完成；有些鱼类和两栖类在小宝宝出生后，妈妈就离开了，爸爸"既当爹又当妈"，比如达尔文蛙，爸爸会把受精卵含在嘴里，直到小蛙长大后才放出来。

娃娃鱼妈妈在产卵后就完成了所有任务，回到它原来居住的洞穴继续自由自在的单身生活，照顾小娃娃鱼的任务就全部落到了娃娃鱼爸爸的身上。它会把身体弯曲成半月形，把卵带护在身体中间，以免被水流冲走。如果天敌来吃娃娃鱼宝宝，它会张大嘴巴，露出锋利的牙齿以示威胁，如果对方还要靠近，娃娃鱼爸爸就会和对方对峙甚至发生打斗，直到对方离开。这种照顾的行为会一直持续到小娃娃鱼出膜，并且能独立生活，娃娃鱼爸爸才会离开。

娃娃鱼的孵化期有多长

娃娃鱼受精卵的孵化期为30～40天，最多也有长达80天的。孵化速度由水温决定，一般水温越高，速度越快，但是过高的温度也会导致卵内胚胎发育畸形或死亡。

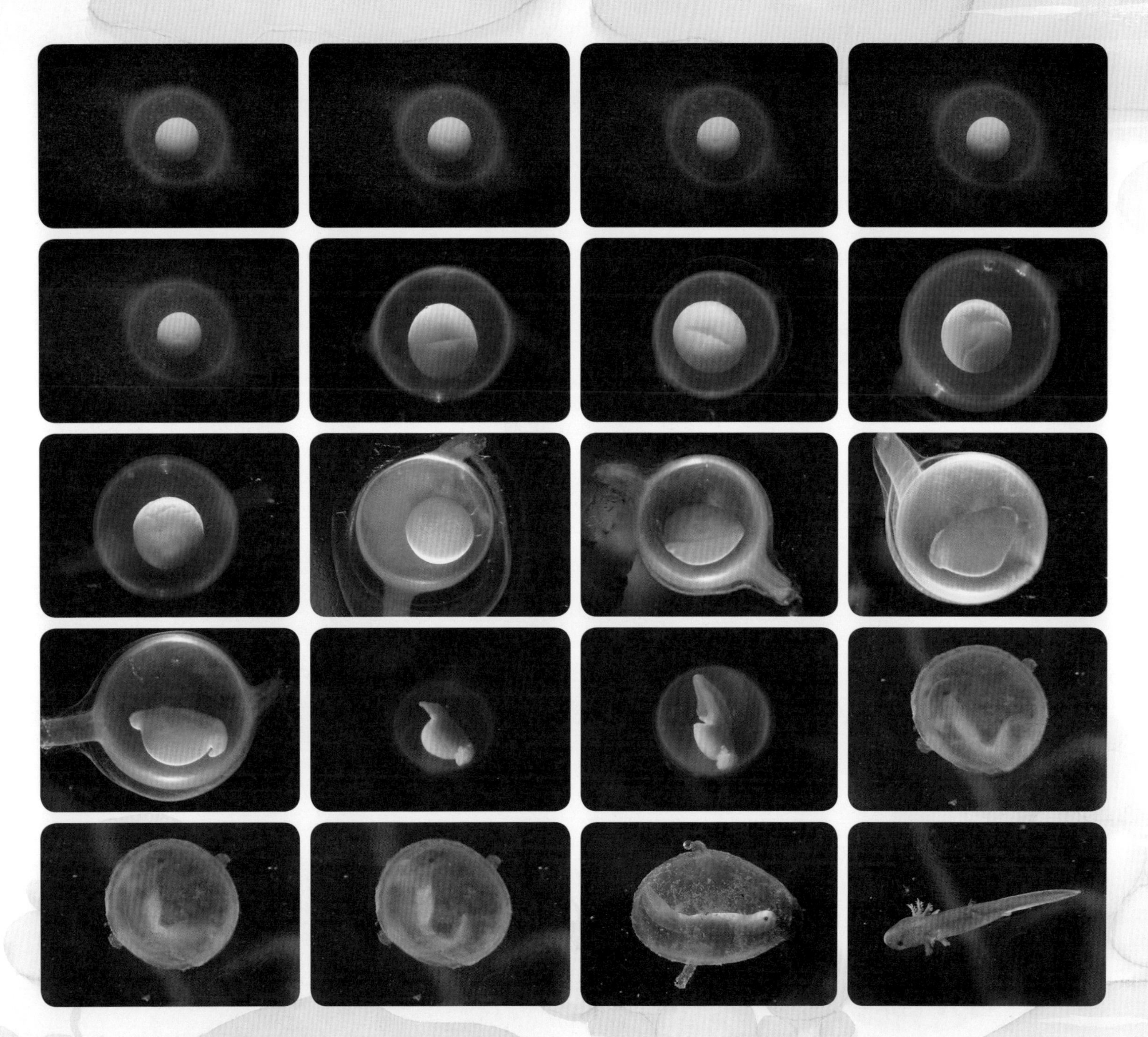

受精卵的生长分为卵裂阶段、囊胚阶段、原肠胚阶段、神经胚阶段、器官形成阶段和孵化阶段。从神经胚阶段开始，可以较为清晰地看见娃娃鱼胚胎成形、各器官慢慢发育完整的过程。在人工养殖条件下，经常可以见到双胞胎甚至三胞胎的情况，在野生情况下，多胞胎的情况不是很常见。

娃娃鱼小时候和小蝌蚪一样吗

两栖动物的成长中有一个必经的过程，这个过程被称为“变态”。因为它们的幼体和成体在形态和生理功能上存在着巨大差异。童话故事《小蝌蚪找妈妈》展现的就是从小蝌蚪到青蛙的这一变化过程。

娃娃鱼和青蛙一样，也是卵生动物。刚刚钻出卵膜的小娃娃鱼没有四肢，脑袋两边各长有三对外鳃，长得很像蝌蚪，体长为 28 ~ 31.5 毫米，全身密布黑色小点。这时小娃娃鱼完全不能离开水，它只能靠这三对外鳃呼吸，每根鳃枝上长着 14 ~ 15 束像绒毛一样的桃红色须状物，上面布满了血管，这是它们幼年期在水中的呼吸器官。过了一段时间，小娃娃鱼会先长出两个前肢，但还没有分叉，后肢的肢芽也会慢慢长出来。小娃娃鱼还没有发育出作为运动器官的前后肢，只能依靠发达的尾部摆动进行不规则的运动，在水里还不能保持平衡，大多数时间都是侧卧在水底休息。再过段时间，小娃娃鱼的后肢长出来了，形成了四叉五趾，就预示着它所有的生理器官都已经形成并且可以成熟运作了。刚刚出生的小娃娃鱼肚子是长椭圆形的，占全身的比例较大，里面含有袋状的卵黄囊，囊内积存的卵黄物质是小娃娃鱼出生后的主要营养来源，直到后肢芽长出以后才被完全吸收消失。它的血液颜色从粉红色慢慢变为鲜红色，皮肤也逐渐变深，有的开始出现斑块。三对外鳃会从外到内逐渐萎缩脱落，皮肤也开始分泌黏液。小娃娃鱼生长非常缓慢，9个月以后才能完成全身器官发育，可以逐步爬出水面生活，并开始用肺呼吸。

娃娃鱼能离开水吗

虽然娃娃鱼可以用肺呼吸，但其实它的肺进化得并不完善，肺泡的表面积比较小，并不足以支持它完成全部的呼吸。而娃娃鱼的皮肤里有丰富的血管和腺体可以进行气体交换，身体旁边的褶皱更是增加了皮肤表面吸收氧气的面积，因此它可以通过湿润的皮肤辅助呼吸。这个原因也决定了虽然娃娃鱼可以上岸，但它必须生活在水中。

娃娃鱼喜欢住在哪里

野生娃娃鱼的栖息地，都是山林茂密、水源充足、空气新鲜而湿润的天然氧吧，那里海拔高度200～1 000米，以300～800米居多，植被覆盖率高达80%～90%。因为娃娃鱼的肺部进化不完全，需要靠湿润的皮肤辅助呼吸，茂密的丛林既可以遮挡阳光的直射，又营造了湿度较大的空气环境，避免它因为皮肤干燥而窒息。流动的溪水是野生娃娃鱼生存的基本条件之一。

温度对娃娃鱼的生长起到了相当重要的作用，娃娃鱼喜欢在16～28℃的水温下活动，而当水温低于4℃或高于33℃时，它们的摄食量就会减少，行动也趋于迟钝，生长非常缓慢。

娃娃鱼对水质有要求吗

可以说娃娃鱼有“洁癖”，它对水质的要求非常高。溶氧量常常作为水质检测的重要指标之一，是指水中氧气的溶解量，也是水体自净能力的标识。野生娃娃鱼生活的水质必须清爽洁净且富含矿物质，水中的溶氧量不能低于5毫克／升。水中生活的生物依靠溶解在水中的氧气呼吸生存。越是干净的水，所含溶解氧越多；水污染越厉害，溶解氧就越少。当溶氧量低于4毫克／升时，就会引起鱼类窒息死亡，有些生活在水流湍急流域的鱼类甚至需要12毫克／升以上的溶氧量。对于人类来说，健康的饮用水中溶氧量不得低于6毫克／升。而对于娃娃鱼，当水中的溶氧量在5毫克／升以上时，清爽无污染的水质最适合它的生长发育。尤其是在幼体阶段和繁殖阶段，水中的溶氧量必须保持在5.5毫克／升以上，其净度已经接近饮用水的标准了。

娃娃鱼是怎么挑选洞穴的

为了尽量避免阳光照射，有个安静的栖息场所，娃娃鱼对洞穴的选择也很有讲究，喜欢栖息在石灰岩地区。在河水长期的侵蚀下，这些地区会产生许多有回流水的自然溶洞、暗河、洞隙和洞穴，成为娃娃鱼栖居的家园。这些洞穴不但为娃娃鱼提供了避光休息的家园、躲避天敌捕食的避难所，而且因富含多种水生生物，还成为娃娃鱼绝好的捕猎场。一旦认准了洞穴，娃娃鱼轻易是不会搬家的。

娃娃鱼是近视眼吗

世界上的大部分两栖动物，以及一部分爬行动物、哺乳动物和昆虫属于夜行性动物，这些动物选择夜晚外出活动的重要原因之一就是为了避开日间强烈的阳光，娃娃鱼也是如此。娃娃鱼的眼睛不像人类的眼睛，当我们感觉光线很强的时候，我们会闭上眼睛来保护我们的眼球。娃娃鱼的眼睛没有眼睑，所以它们不能眨眼睛，因此非常害怕强光的刺激。

娃娃鱼的眼睛结构比较简单和原始，眼睛里的细胞分布决定了它以夜行性为主，它的视觉分辨率很低，只能看到模糊的影像。上亿年来，它们只习惯于洞穴里非常昏暗的光线，一对小眼睛逐步退化成了高度近视眼，这也大大增加了它们在白天活动的危险性。

娃娃鱼喜欢吃什么

每当夜深人静，就是娃娃鱼出洞活动、补充能量的时候。娃娃鱼的食物来源比较丰富，2岁以内的小娃娃鱼主要以植物为食，幼年的娃娃鱼会吃一些无脊椎的水生动物，比如螃蟹、虾，以及昆虫的幼虫等。成年娃娃鱼的菜单更加丰盛，从鱼、青蛙、水蛇、水鼠等水生脊椎动物，到螺、蚌等软体动物，甚至还会吃动物的尸体。

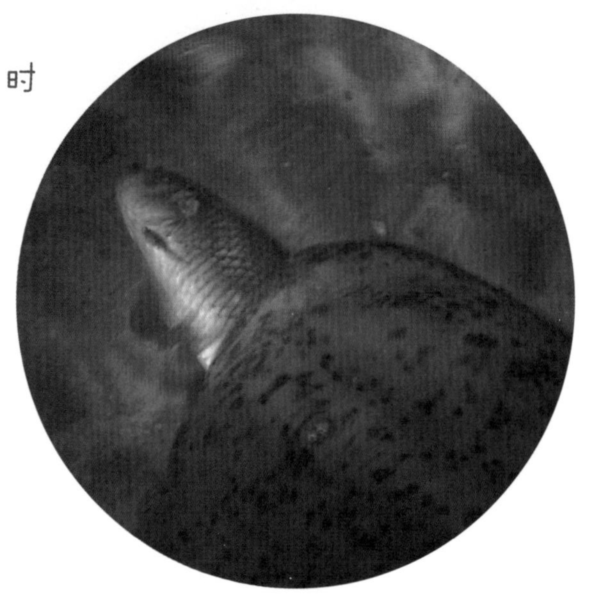

娃娃鱼很懒惰吗

成年娃娃鱼的行动随着年龄的增长变得迟缓笨拙，远远比不上幼年时期行动灵活敏捷，所以它们并不擅长追捕猎物。那么娃娃鱼是怎么捕食的呢？有一句形容好吃懒做之人的话叫做“衣来伸手，饭来张口”，娃娃鱼就是典型的“饭来张口”。傍晚的时候，娃娃鱼会盘踞在洞口附近水滩的乱石丛中，用和自己身体颜色相近的石块作掩护，借着夜色隐匿起来，当猎物经过时，它就突然发动袭击，用尖利的牙齿死死咬住猎物不放。猎物一旦被咬住，就很难逃脱。正是因为娃娃鱼这种看似被动、实则高明的猎食手段，在民间就流传着“娃娃鱼坐滩头，喜吃自来食”的谚语。

娃娃鱼的牙齿厉害吗

娃娃鱼的上颚有两排牙齿，下颚有一排牙齿，都非常细密尖锐，上下颚合拢时牙齿便形成交错状。哪怕猎物挣扎，它也能紧紧咬住不松口。但是，这些细密尖锐的牙齿却不能咀嚼，所有的食物都被娃娃鱼直接囫囵吞下，然后在胃里慢慢消化。娃娃鱼非常耐饿，可以很长时间不吃东西，因为它的新陈代谢很慢。但是它吃起来，胃口可以很大，甚至到达暴食的程度。它可以一口吞下自己个头一半大的猎物，如果食物比较多，一顿饭就可以增加自身体重的五分之一。因此，有人将娃娃鱼的胃比喻成“铜墙铁壁”，来赞叹它超强的消化能力。

娃娃鱼有天敌吗

成年的娃娃鱼处于水下食物链的顶端，它在水中几乎没有敌手。在自然界中的敌人，是黄鼠狼、野猫、水獭等夜间出没的小型食肉动物。一旦不幸遇敌，它就用锋利的牙齿和有力的尾巴进行自卫。如果还不能脱身，它就把胃里的臭鱼烂虾朝敌人吐去，把对手吓跑。争斗更加激烈的情况下，如果被对手咬住，实在脱不了身，它还有最后一个保命的绝招，它会从颈部毛孔分泌出一些黏糊糊的白色毒液。娃娃鱼分泌的黏液有一种强烈的花椒味，刺激对手被迫离开。黏液本身的毒性并不强，但是黏着力非常厉害，往往会粘住对手的口唇，最后只好眼睁睁地看着它成功脱逃。日本大鲵俗称“大山椒鱼”，也正是因为身上有这种黏液的缘故。

哪里曾经发现过娃娃鱼

娃娃鱼曾广泛分布于我国长江、黄河和珠江流域17个省区市的偏远山区的溪流中，这17个省区市分别为河南、陕西、山西、甘肃、青海、四川、重庆、贵州、湖北、安徽、浙江、江西、湖南、福建、广东、广西和云南。

娃娃鱼主要集中分布在四大区域：一是湖南张家界、江永、岳阳和湘西自治州；二是湖北房县、神农架；三是陕西安康、汉中、商洛；四是贵州遵义和四川宜宾、文兴等地。曾是我国分布范围最广的有尾目两栖动物。

在20世纪80年代以前，野生娃娃鱼的种群数量非常多。如果以20世纪60年代娃娃鱼存活量为基准（100%）的话，21世纪中国大鲵的存活量只有当时的10%～15%。2000年我国野生娃娃鱼资源量仅为5万～20万尾，只相当于1980年全国存活量的2%。

中国大鲵长期以来被认为是一个物种，但根据科学家最新的DNA分析结果，可能是由于大鲵种群在百万年前被隆起的山脉彼此隔离，从而形成至少五个物种。而根据科学家在2013～2016年进行的实地调查中，在16个省的97个分布点总计才发现24条野生娃娃鱼，野生娃娃鱼的现状可能比我们想象的还要岌岌可危！